PREDATORS

TARANTULAS

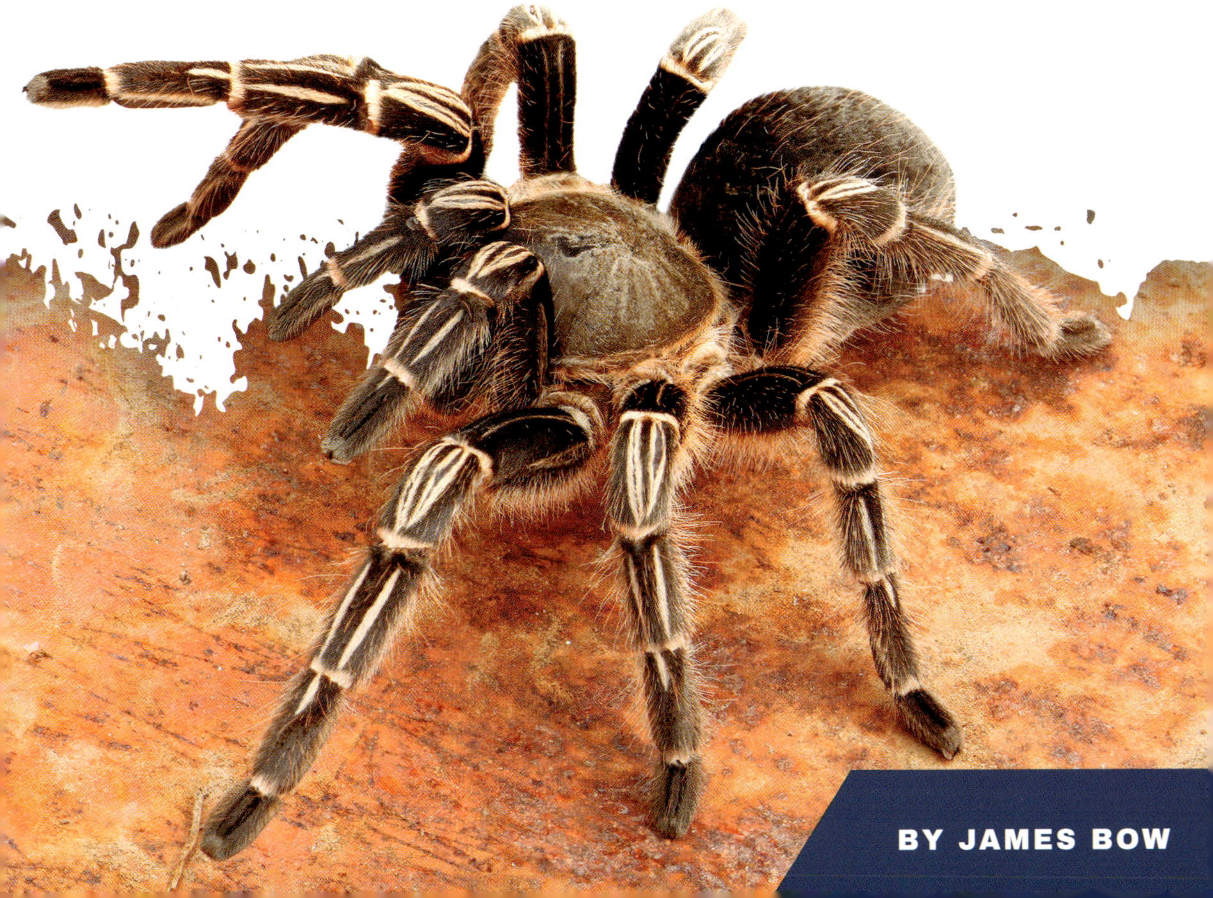

BY JAMES BOW

WWW.APEXEDITIONS.COM

Apex is distributed by North Star Editions:
sales@northstareditions.com | 888-417-0195

Produced for Apex by Red Line Editorial.

Photographs ©: Shutterstock Images, cover, 1, 4–5, 6–7, 8–9, 10–11, 12–13, 14, 15, 16–17, 18, 19, 20–21, 22–23, 24, 25, 26, 29

Library of Congress Control Number: 2023910176

ISBN
978-1-63738-777-1 (hardcover)
978-1-63738-820-4 (paperback)
978-1-63738-901-0 (online PDF)
978-1-63738-863-1 (hosted ebook)

Printed in the United States of America
Mankato, MN
012024

NOTE TO PARENTS AND EDUCATORS

Apex books are designed to build literacy skills in striving readers. Exciting, high-interest content attracts and holds readers' attention. The text is carefully leveled to allow students to achieve success quickly. Additional features, such as bolded glossary words for difficult terms, help build comprehension.

TABLE OF CONTENTS

CHAPTER 1

SURPRISE ATTACK

A tarantula sits near its **burrow**. The large spider holds very still. The hairs on its legs sense movement. A mouse is walking nearby.

Some hairs on a tarantula's body sense movement in the air. Other hairs help the spider taste.

Tarantulas grab animals with their front legs before they bite.

The tarantula waits for the mouse to come close. Then it jumps out and bites.

FAST FACT

Tarantulas cannot see well. They rely on their sense of touch to find things.

The tarantula grabs the mouse in its strong jaws. Its fangs release **venom**. The mouse dies, and the tarantula eats it.

A tarantula uses its sharp fangs to poke holes in its prey.

QUICK STRIKE

Tarantulas do not chase their **prey**. They stay still and wait to attack until prey comes close. Animals that hunt this way are called ambush predators.

CHAPTER 2

STUNNING SPIDERS

There are hundreds of different types of tarantulas. They come in many colors and sizes. But all have hairy bodies.

Like all spiders, tarantulas have eight eyes and eight legs.

Goliath bird-eating spiders are large enough to eat scorpions.

The largest tarantula is the goliath bird-eating spider. It can grow up to 11 inches (28 cm) long. The spruce-fir moss spider is the smallest tarantula. It's about the size of a pea.

FAST FACT

Tarantulas can hiss when **threatened**. They do this by rubbing the hairs on their legs together.

Some types of tarantulas use holes in trees as their burrows.

Tarantulas can be found on every continent except Antarctica. But most live in Central or South America. They make their homes in warm places.

TARANTULA HOMES

Many types of tarantulas dig burrows in dirt or sand. Some live under rocks or logs. Tarantulas may also live in trees.

Tarantulas often live in dry areas such as deserts or grasslands. But some types also live in rainforests.

CHAPTER 3

LIFE CYCLE

Most tarantulas live alone. However, they come together to **mate** once a year. Later, females lay eggs. The females make sacs to hold the eggs.

Each egg sac can hold hundreds of eggs.

Spiderlings start to hunt when they are about a week old.

Females guard their eggs until they hatch. Baby tarantulas are called spiderlings. They can live on their own soon after hatching.

SPINNING SILK

Tarantulas spin silk to make egg sacs. Some tarantulas also use silk to line their burrows. Or they may make trip wires. These pieces of silk warn the spiders when other animals come near their burrows.

Greenbottle blue tarantulas often live in tunnels lined with silk.

Tarantulas molt each time they grow. They shed their **exoskeletons** and grow new, bigger ones.

FAST FACT

In the wild, most male tarantulas live for about seven years. Females can live to be 30.

Young tarantulas molt about once a month.

CHAPTER 4

NIGHT HUNTER

Tarantulas usually hunt at night. They mostly eat insects or small spiders. Larger tarantulas can also eat lizards, snakes, frogs, or birds.

Tarantulas are carnivores. That means they eat meat.

Some tarantulas' fangs can grow 1 inch (2.5 cm) long.

Tarantulas bite prey with hollow fangs. The fangs **inject** venom. It **paralyzes** the prey.

HAIR ATTACK

Some types of tarantulas have sharp, stinging hairs on their backs. The spiders use these hairs for defense. They use their back legs to flick the hairs at an attacker's face.

Zebra tarantulas are one type that can flick stinging hairs.

Next, tarantulas spit **digestive** juices onto the prey. The juices turn the inside of the prey's body to liquid. Tarantulas suck up this liquid.

Tarantulas sometimes bring prey back to their burrows to eat.

COMPREHENSION QUESTIONS

Write your answers on a separate piece of paper.

1. Write a few sentences describing how tarantulas hunt.

2. Would you like to keep a tarantula as a pet? Why or why not?

3. What is the largest kind of tarantula?

   A. spiderling
   B. spruce-fir moss spider
   C. goliath bird-eating spider

4. How might a tarantula use a trip wire?

   A. to cover its eggs
   B. to sense when prey comes near
   C. to fight back if attacked

5. What does **molt** mean in this book?

*Tarantulas **molt** each time they grow. They shed their exoskeletons and grow new, bigger ones.*

A. lose and replace an outer layer
B. gain a larger number of teeth
C. have a smaller number of legs

6. What does **defense** mean in this book?

*The spiders use these hairs for **defense**. They use their back legs to flick the hairs at an attacker's face.*

A. a way to make noise
B. a way to fight back
C. a way to run fast

Answer key on page 32.

GLOSSARY

burrow

A tunnel or hole an animal uses as a home.

digestive

Related to breaking down food so the body can get energy from it.

exoskeletons

Hard coverings that protect the body.

inject

To send something into an animal's body.

mate

To form a pair and come together to have babies.

paralyzes

Makes something unable to move.

prey

Animals that are hunted and eaten by other animals.

threatened

Put in danger.

venom

A poison made by an animal and used to bite or sting prey.

BOOKS

Adamson, Thomas K. *Scorpion vs. Tarantula*. Minneapolis: Bellwether Media, 2021.

Gitlin, Marty. *Tarantulas*. Mankato, MN: Black Rabbit Books, 2020.

Jaycox, Jaclyn. *Tarantulas*. North Mankato, MN: Capstone Publishing, 2020.

ONLINE RESOURCES

Visit **www.apexeditions.com** to find links and resources related to this title.

ABOUT THE AUTHOR

James Bow is the author of more than 75 educational books for kids. He lives with his wife, two kids, two cats, and one dog in Kitchener, Ontario, Canada. He likes learning about tarantulas but would never want one as a pet.

INDEX

ANSWER KEY:
1. Answers will vary; 2. Answers will vary; 3. C; 4. B; 5. A; 6. B